CAFFEINE AND NICOTINE

Many teens don't realize the harmful effects of caffeine and nicotine.

CAFFEINE AND NICOTINE

Richard S. Lee and Mary Price Lee

A HAZELDEN / ROSEN Book

To Margo and Tony: "Sister" and "Brother"

First published in 1994 by The Rosen Publishing Group, Inc.
29 East 21st Street, New York, NY 10010

This edition published in 1997 by Hazelden
PO Box 176, Center City MN 55012-0176

First Edition

Library of Congress Cataloging-in-Publication Data

Lee, Richard S.
    Caffeine and nicotine / Richard S. Lee and Mary
  Price Lee.
        p.   cm.—(The Drug abuse prevention library)
    Includes bibliographical references and index.
    ISBN 1-56838-168-9
    1. Caffeine—Juvenile literature. 2. Caffeine
  habit—Juvenile literature. 3. Nicotine—Juvenile
  literature. 4. Tobacco habit—Juvenile literature.
  5. Drug abuse—Prevention—Juvenile literature.
  [1. Caffeine. 2. Nicotine. 3. Drugs. 4. Drug
  abuse.] I. Lee, Mary Price. II. Title. III. Series.
  HV5809.5.L44 1994
  613.85—dc20   94-2279
                                                                         CIP
                                                                          AC

Manufactured in the United States of America

Contents

Introduction

*D*id you know that when you drink a cola or eat a chocolate bar, you are using a *drug*? That drug is completely legal for people of any age to buy, but it may be harmful you.

This book is about two drugs. You can become addicted to them; that is, your body can crave them. They can harm you, and their dangers are well known.

These drugs are *caffeine* and *nicotine*. You should watch carefully how much you get of the first one, if you feel that you must use it. You should not use the second one at all.

This book will tell you why these drugs are dangerous.

Why Are They Sold?

Many people still use caffeine and nicotine because they don't want to believe the drugs are harmful. Caffeine, a pepper-upper, was not always taken seriously. It was—and is—found in coffee, tea, chocolate, candy bars, cola drinks, stay-awake medicines, cold tablets, and pain relievers. Most people thought it was harmless. But other people complained about "coffee nerves" and "cola jitters." Scientists developed decaffeinated coffees and teas and caffeine-free soft drinks to solve the problem.

Nicotine is a truly dangerous drug. It is in all tobacco products—cigarettes, cigars, pipe tobacco, chewing tobacco, and snuff. Snuff was once inhaled. It is now called "smokeless tobacco" and placed in the mouth. People have used tobacco for hundreds of years. Only now do we know that tobacco causes many extremely serious health problems.

"Twenty Hits of Nicotine, Please"

No one would ask for cigarettes that way, but 20 doses of the drug nicotine are what you get when you buy a pack of cigarettes. The packages do not list the

Nicotine addiction is the number one cause of death in America.

ingredients, but they do now carry health warnings.

Did you know that 50 million Americans, or 26 percent of the population, use nicotine, mostly in cigarettes? Did you know that nicotine addiction *kills* about 390,000 people every year? It is the number one cause of death in America. (In comparison, a major "hard" drug, heroin, kills 3,000 people a year.)

Caffeine also can cause problems if you take too much. It can be addictive, too. Your body may want more and more, especially when the caffeine in your body wears off. Then you can feel let down and sleepy. You can even become depressed.

Experts don't agree on which are the best ways of getting unhooked from nicotine addiction. Would being hypnotized work for you? Would the ancient Chinese art of acupuncture get you unhooked? Should you join a support group of people who are trying to quit smoking? Should you see a doctor for a nicotine "patch"?

You may need to get unhooked from caffeine, too. Suppose you are a "soda-head," drinking four, five, or up to nine or ten cans of caffeinated soda every day. Does this give you too much caffeine? Do you drink coffee? How many cups a day? How can you be hurt by the "stay-awake" pills? What could happen to you by taking too much caffeine?

Read on and find out how to cope with caffeine and nicotine.

You can become addicted to caffeine by drinking as few as two colas a day.

CHAPTER 1

What Is Chemical Dependency?

Chemical dependency simply means "getting hooked." Here's how it happens. We all have receptors or pleasure centers in our brain. These pleasures centers act the way our stomach does when it tells our brain that we're hungry. Your pleasure centers can demand action, just like your stomach.

This can happen when your pleasure centers are exposed to the chemicals in certain drugs. These drugs change the way you feel, see things, or react. Some drugs make you more alert, or give you a false feeling of energy. Others can take you away from real life and change your reactions to everyday situations. Some of these changes are dangerous. Many drugs

12 | are unsafe even in small doses. Most of these are illegal or "street" drugs.

Caffeine and nicotine are two chemicals that can make your pleasure centers want more. (Alcohol is another. And chemicals in other drugs can have similar effects.) Addiction takes place when the pleasure centers take over and demand what they want, whether you like it or not.

The Good News
Caffeine and nicotine are sold legally in products of controlled quality. You can't accidentally overdose on them. Smoking can make you dizzy or sick if you're not used to it, and too much caffeine can make you feel jittery. But you only have to watch out for the *quantity* you absorb, not the quality.

The Bad News
Both caffeine and nicotine are *addictive*. Little by little, your pleasure centers will demand more of the products that contain these drugs. And the more the pleasure centers demand, the more addicted you can become.

A Walk in the Sun
It is Senior Week, party time at the New Jersey seashore. Aaron and Giorgio, along

Joe Camel is nearly as well known to kids under six as Mickey Mouse.

14 | *with several other members of their high school graduating class, are staying in Wildwood. On this bright morning, they go to the boardwalk for a stroll by the ocean.*

"Is he big, or what?" Aaron says, pointing at a billboard of Old Joe Camel shooting pool with his animal friends.

"Not as big as the other guy," replies Giorgio, looking back down the boardwalk. Aaron turns to follow his gaze. Although it is a misty morning, they can both see the Marlboro Man on a huge billboard several blocks away.

"They're trying to tell us something," Aaron says.

"You bet they are," says Giorgio. "Take your choice. Joe Camel says, 'Come into my cartoon world of smooth characters if you want to have fun.' The Marlboro Man says, 'It's macho to smoke—part of the great American outdoors.' "

"That's not what the real Marlboro Man said before he died of lung cancer," Aaron recalls. "Wayne McLaren said not to smoke, remember?"

"I sure do," says Giorgio. "But nobody hears that message when the tobacco companies cover the landscape with billboards. They want us to smoke."

"Well," says Aaron, *"they have to get their* *three thousand new smokers a day from somewhere!"*

Joe Camel and Mickey Mouse

Cigarette advertising is seen by millions of people under age 18, the legal age for tobacco use. This can't all be accidental. At least two tobacco companies use cartoon characters that naturally appeal to children. A recent survey showed that Joe Camel was almost as well known to kids under six as Mickey Mouse.

Some cigarette companies put fake money in cigarette packs, to be used to "buy" sweatshirts, caps, and jackets with the brand name on them. To get these things, you have to sign a statement that you are 21 years old. But the very hip products are aimed at the people who want the "in" feeling this kind of merchandise gives them.

Although cigarette companies are no longer allowed to advertise on television, they spend millions of dollars to promote televised sports events that young viewers enjoy.

The events are named for cigarettes. So the brand names are mentioned on the air. Brand-name banners and signs

16 and name-painted racing cars are seen on the screen.

Do the race drivers, tennis stars, and others smoke the cigarette brands that sponsor their events? Few of these sports figures smoke at all. They know that smoking would hurt their performance. So why do the tobacco companies do so much promotion involving sports stars and television?

The answer is simple: *brand awareness.* The more places the product name appears, the better it will be remembered by people. And the better it is remembered, the more it will be used. That's the plan. And it works. Television is our biggest way to reach people. Cigarette makers are determined to be on TV one way or another.

Caffeine and Reality

The makers of caffeinated products, especially cola-type soft drinks, do the same kind of promotion. Tobacco companies base their appeals on taste, satisfaction, and the supposed "safety" of low-tar, low-nicotine tobacco. The makers of caffeinated products say it differently.

Coffee and tea companies do not aim especially at younger users. Their ads talk about starting the day right, refreshment, flavor, and energy. But soft drinks *are*

You can avoid addiction to caffeine by choosing other beverages such as water or juice.

flavor, and energy. But soft drinks *are* aimed at young consumers, since drinks with caffeine are legal at any age. Cola ads say more about being part of the crowd and having fun than about taste, satisfaction, or quenching thirst. Many of their TV ads depend on lavish commercials with entertainers who appeal to young people. They use slogans that make their products sound important.

Health warnings are placed in tobacco advertising because government regulations now require it. But the makers of caffeinated products never mention the possible dangers of caffeine. Products

18 | or "no caffeine" in ads without saying what problems caffeine could cause. (Can you imagine a soft-drink campaign with the slogan, "Drink Alert-Cola and get wired!" As you'll see, one drink comes pretty close.)

What's the Hook?

The "hook" is *chemical dependency,* the addictive nature of caffeine and nicotine. Chemical dependency is why the marketers aim at younger users. They figure if they can get you to drink coffee and cola and to smoke when you're young, they've got you hooked forever. And it's true! Any addiction is hard to break.

How do you cope with these pressures? Enjoy the fancy soft-drink commercials. Smile at Joe Camel and his cast of smooth characters. Watch the Virginia Slims tennis matches and the Marlboro and Winston auto racing teams. Let TV give you a free ride. But *don't buy the products!*

That's easy to say.

It may not be easy to do. The chapters that follow give you the reasons why you can get hooked, and why it is important to control caffeine and avoid nicotine. They will give you tips that can help you stay free of them, or get free of them.

CHAPTER 2

What's Wrong with Caffeine?

*L*egend says that shepherds at a monastery in Arabia were surprised to find their goats jumping around and playing late at night, when they should have been asleep. The shepherds found that the goats had eaten berries from coffee plants. The wide-awake goats were getting a kick from caffeine! Later, the monks (and probably the shepherds) figured out how to brew the beans into coffee and get the same effect.

Experts say that caffeine is the most widely used drug in the world. Most of us associate caffeine with coffee—from which its name came. Between 82 percent and 92 percent of Americans drink coffee—to the tune of about 800 cups a

Many parents set a bad example by drinking caffeinated coffee every morning.

year each! Coffee is probably the easiest way to get large doses of caffeine, but it is not the only way.

Caffeine is an odorless, rather bitter drug found in coffee beans, tea leaves, cocoa beans, and cola nuts. From these sources, caffeine finds its way into many products besides coffee and tea.

Druggists classify caffeine as an *analeptic*, a fancy word for "stimulant." It is also *psychoactive*, which means that it affects your central nervous system. Caffeine, even in small doses, makes your heart beat faster. It makes you more alert. It can help you think more clearly. It may even give you a little more "energy" for running or other sports. What's wrong with that?

The Downside of Caffeine

- The stimulating effects of caffeine can be addictive. Many people say they "just can't get started" in the morning without the caffeine in coffee or tea.
- Caffeine's "lift" is false and does not last long. When it wears off, you may have just the opposite effect. You can feel tired and let down. You want

22 more caffeine to get back the "kick" it gave you.

- You can develop a tolerance for caffeine. Your body will need more and more of it to feel the same lift you once got from just a little.
- Caffeine can make your blood vessels smaller. This makes it harder for your heart to pump blood through your body. So your heart beats faster.
- In larger doses, caffeine can make you anxious and nervous. It can cause stomach problems by creating more stomach acid. It can interfere with your sleep, even several hours after you drink it. It makes you need to urinate more often, and it can give you diarrhea. It can bring on a very scary condition called *tachycardia*, or "runaway heartbeat." Large doses have been banned for athletes in the Olympic Games.
- Some people are so allergic to caffeine that even a little can cause discomfort.
- Some people who drink as little as two cups of coffee a day and then stop often feel withdrawal symptoms. These start in 12 to 24 hours. They

Caffeine is a legal drug, but the amounts in products are regulated by the government.

are at their worst in one to two days and can last for a week. The symptoms include headache (reported by 52 percent in a recent study), less energy (11 percent), depression (11 percent), anxiety (8 percent), and fatigue or drowsiness (8 percent). These symptoms can be severe. They may be one reason why caffeine

24 | users—especially coffee drinkers—do not stop.

Sandy's Day at Home

"Hi, Mom," says Sandy, coming into the kitchen.

"You're not dressed!"

"I'm not going to school. My cold is worse. I feel lousy. I'm going back to bed."

"I fixed your cereal. Do you want an egg?"

"No, thanks, Mom. Cereal's fine. Juice. And I'll get some coffee."

"Just one mug, though."

"No problem. Do we have any pain relievers?"

"I think so. Well, I'm off to work. Will you be OK by yourself?"

"Don't worry about me, Mom. I'll be fine."

Sandy's mother leaves, and Sandy drinks her coffee. She phones her best friend to say she won't be in school. She takes two pain relievers and goes back to bed. She wants to go to sleep, but she can't. So she watches TV all morning.

At lunchtime, Sandy has a peanut butter and jelly sandwich with a cola, and treats herself to the last chocolate bar in the house. Then she takes two more pain pills.

She really feels out of it—jumpy, restless, and a little sick to her stomach. "That's what colds will do," she says to herself.

How Much Caffeine Is Too Much?

Sandy's feeling "out of it" was probably her cold *and* too much caffeine in a short time. Caffeine dosages are measured in milligrams. Here's what she took in just five hours:

One 8-ounce mug of coffee	175 milligrams
Four pain relievers	260 milligrams
One 12-ounce cola drink	50 milligrams
One chocolate bar	20 milligrams
Total	505 milligrams

Doctors say that *adults should not have more than 300 milligrams of caffeine a day.* And people your age should have much less. You may not have reached your adult weight, so you should not bombard a smaller, lighter body with adult amounts of caffeine.

As Sandy proved, without knowing it, her half-day at home gave her three to four times the caffeine she should have had. Unless you know everything that

26 | contains caffeine, it's hard to control your intake.

Caffeine is found in surprisingly many products. The dosages (mg) in the chart on page 27 are average:

If you eat sugary foods like chocolate cake or candy bars, put sugar in coffee or tea, or drink regular cola drinks, you get an extra lift from the sugar. You also get an extra let-down when the "sugar rush" and the "caffeine kick" both end.

You can *really* get too much caffeine in a hurry if one of your choices is Jolt Cola. Its slogan is: "All the sugar and twice the caffeine." Jolt contains 5.9 milligrams of caffeine per ounce, just within the safe limit set by the U.S. Food and Drug Administration.

Many other cola drinks contain more caffeine than is in their cola beans. Some soft-drink makers actually buy the caffeine removed from decaffeinated coffee and add it to their products. The caffeine in some medicines and pain relievers comes from decaf coffee, too.

How Caffeine Works

The caffeine you take in goes to work within 15 to 45 minutes. It stays on the job for from 2 1/2 to 7 1/2 hours. You

Food or Product	Caffeine in Milligrams
Coffee, 8 ounces	
Drip grind	175 mg
Percolated	135 mg
Instant	100 mg
Decaffeinated	4 mg
Tea, 5 ounces	
Tea bag or leaf	40 mg
Instant	30 mg
Cocoa, 8 ounces	7 mg
Chocolate, solid, 1 ounce	20 mg*
Soft drinks, 12 oz.	35 mg (Coca-Cola)
	52 mg (Mountain Dew)
	71 mg (Jolt Cola)
Stay-awake pills, each	100 mg (NoDoz)
	200 mg (Vivarin)
Pain relievers, each	32 mg (Anacin)
	65 mg (Excedrin)
Diet/weight-loss pills, diuretics (water removal pills), each	100–200 mg

* Many candy bars contain more caffeine per serving than shown for plain chocolate.

28 | won't feel its effects after that—but it can take up to 30 hours for your body to get rid of it.

The "kick" in caffeine comes from molecules similar to those of *adenosine,* a chemical that your body manufactures. Adenosine keeps your brain slowed down. It controls brain activity through *receptors,* or gateways. When you drink a beverage containing caffeine, the brain receptors take the caffeine molecules for adenosine. The receptors "link" with the caffeine and block out the adenosine. Without adenosine to slow it down, your brain revs up. If you keep on drinking caffeine, your body supplies more and more receptors. You have to drink more and more caffeine to stay alert.

Does Coffee Really Act Like a Drug?

Luis Oswaldo Rodrigues, a noted professor, says yes. He conducted research on coffee-drinking athletes who consumed much less caffeine than the 750-milligram limit set by the International Olympic Committee. Dr. Rodrigues found that the long-distance runners increased their performance up to 20 percent with the caffeine and ". . . were under a doping effect."

By exercising and eating healthy foods, you keep your body in the best possible shape.

How to Manage Caffeine

First, don't say it won't affect you. Caffeine affects *anyone* who takes it. Since it is absorbed by the body very quickly, the effects are almost immediate.

It affects some people more than others. The caffeine that gives one person a mild lift makes another person jittery and nervous, increases the heartbeat, or makes him irritable.

How do you manage caffeine?

- Decide *not* to use it. If you have been using it, gradually reduce what you

30

take until you're caffeine-free. This will help you avoid the discomfort of withdrawal symptoms.

- If you must use caffeine, decide to *control* what you take. Figure out how much that is, and where it comes from. If you're getting more than one milligram per day for every pound you weigh, cut down—you are at the max! Reduce your intake over several days. Then stick with your plan.

- Plan *substitutes*. Decaffeinated coffee, decaf or herbal teas, caffeine-free soft drinks are all easy to find. Or drink healthful alternatives such as fruit juices or fruit-flavored "all-natural" sodas. Cut back on chocolate, cocoa, and things made with them. Use carob products instead of chocolate—the taste is almost the same. Substitute hard candy for chocolate candy bars. Avoid pain relievers (use plain aspirin instead)—and *stay away from "stay-awake" pills!*

Study Jitters

"Man, if she gives us an essay question, I'm sunk!" says Diana. "What time is it?"

"Twelve-thirty," Eric answers. "We've

Natural foods and beverages give your body the energy it needs to be its best.

been cramming since four this afternoon! I've had it! I'm so jumpy, I could party all night—but I can't get this stuff to stick in my head. I don't think I'll remember anything in the morning."

"Me, either. But now I can't slow down enough to sleep," Diana says.

"Maybe we shouldn't have taken those pills."

Diana and her brother made a common mistake. First, they drank caffeinated cola when they started studying. Then they figured that if one stay-awake pill would

Too much caffeine can leave you sleepless at night.

help them concentrate, two would help them concentrate *better*. What they forgot was that the soda plus the pills affected them like four cups of coffee: They had real jitters and lost their ability to think straight. As for getting a good night's sleep? Forget it!

This next event really happened. It was described in *The Washingtonian* magazine. A doctor told about a student who was "pulling an all-nighter" to study and thought he was having a heart attack. He drank five colas, eight cups of coffee, and

took eight stay-awake pills. By 2:30 in the |
morning, he had cramps and diarrhea. By
3:00 he had the sweats and an irregular
heartbeat. He drove himself to a hospital
emergency room. The diagnosis was
"study anxiety."

It was really caffeine overdose. If you
add it up, using the chart on page 27,
you'll see that he had taken in 2,450
milligrams of caffeine—a daily dosage
safe only for someone weighing more
than a ton!

What's Wrong with Nicotine?

*I*f you watch old movies on TV, you see lots of smoking. Years ago, no one knew that smoking was harmful. People thought it calmed your nerves and made hard situations easier. That's why half of all Americans smoked. Now only about a quarter of them smoke. Many smokers still believe that smoking calms them down, but it is not true. The belief is one reason it's so hard to quit smoking.

Nicotine is a *stimulant*. It speeds up the way your body and brain work. It makes your heart beat faster. It increases your blood pressure. Some (but not all) smokers feel a "kick" or "lift" when they smoke. Some smokers also claim that smoking helps them relax. In fact, a 1992

Gallup Poll showed that teens listed the calming effect of smoking when they felt stressed as the number 1 reason for smoking.

What actually makes smokers nervous isn't stressful situations but *the body's compulsive need for more nicotine.* The "relaxation" of smoking happens when the smoker lights up and the pleasure centers start getting a nicotine "fix."

Long ago, young people also had other excuses for smoking. Today, they still use the same ones:

- "My friends smoke."
- "I don't want to feel left out."
- "It's cool."
- "My parents smoke, and they're not sick."
- "It bugs grown-ups when I smoke."
- "It keeps me from getting fat."

Why Teenagers Smoke

Would you stick your finger into an electric socket just because someone dared you? Of course not—so why would you say yes when someone dared you to smoke? When young people who smoke try to influence *others* to smoke, it is *peer*

36 | *pressure.* It's the number one reason people start to smoke.

It's also the dumbest. Here's why:

- Three out of four of your peers (those your age) do not smoke. *They* are the "in crowd" worth joining.
- Smoking is no longer cool. Smoking is down by 50 percent over recent years. It is permitted in fewer and fewer places. Growing numbers of people don't allow smoking in their homes. Many office buildings are now entirely smoke-free. Some drugstores have stopped selling any tobacco products.
- Cigarette smoke gets into everything—your clothes, your hair. You have a strong smoke smell, especially to nonsmokers. You also have bad breath.
- Smoking works against your exercise, workout, or sports plans.
- Smoking on school grounds may get you suspended. Do it several times and you might be expelled.

How can you resist the peer pressure to smoke? Here are some ideas:

The easiest way to avoid nicotine addiction is not to try smoking.

- Make it clear to your smoker friends that you don't smoke, period. If they want to do things with you, fine. If they don't, that's fine, too.
- If you've tried smoking and didn't like it, say so. Tell your friends it does nothing for you.
- Don't be influenced by the fact that members of your family smoke. Smoking is their choice. It doesn't have to be yours.
- Let nonsmoker friends know you agree with them. Saying so can help make you one of *their* peer group.

- Do things and meet people unrelated to smoking.

Two Other Excuses for Smoking

Billy and Jack are walking home from basketball practice. Billy lights up a cigarette.

"Hey, I thought you quit smoking," Jack says.

"I can quit any time I want to," says Billy. "It won't hurt me if I smoke in the meantime. And so what? I don't care if I don't get to be old. Who wants to be 85 years old, sleeping all day in a wheelchair?"

"You don't need to be worried about that!"

Jack is right. If his friend keeps on smoking, the odds are he won't make it even close to age 85. He may not see 60. Also, Billy is wrong on two other counts.

- *"I can quit any time I want to."* Many teens—and adults—believe this. They are genuinely surprised to find that their bodies won't let go of nicotine. Surveys of teens who planned to quit smoking show that most had *not* quit one or even two years later.

TEENS AND SMOKING

A survey of 9,965 teens by the National Center for Health Statistics showed that:

- One out of two teens (50 percent) who hang out with smoking friends also begin to smoke. Only 3 percent who have nonsmoker friends start smoking.
- Three out of ten teens whose older brothers or sisters smoke take up the habit. And 15 percent of teens whose parents smoke will smoke, too.
- If you start smoking early, you are 16 times more likely to smoke as an adult than if you started smoking after age 21.
- Teens misjudge the addictive power of cigarettes. Of the 3.7 million teens who smoke, 92 percent say they don't plan to be smoking in a year. But only 1.5 percent manage to quit!

- *"It won't hurt me if I smoke in the meantime."* Wrong. Smoking can have immediate effects, even if you're used to it enough not to feel light-headed or sick. Your heart rate increases, the level of oxygen in your blood goes down, and you can lose weight.

Billy doesn't care about living to be 85. Maybe you can't even imagine being that old. You may think that anyone with gray hair, or over 45, is "old." Back when life was harder and medicine was not so advanced, people really were "old" at 45. Now, age is a matter of how you feel rather than how many years you've lived. Smoking can make you old before your time. Smokers get wrinkles long before nonsmokers do. And smoking contributes to many killer diseases.

The fact that smoking can keep weight down is one reason many teenage girls smoke. The Centers for Disease Control report that:

- 34 percent of high school girls think they are overweight.
- 44 percent are trying to lose weight.
- Only 20 percent really are over their best weight.

Trying to control your weight by smoking instead of eating endangers your health.

42 | Since being thin is very important to teenage girls, it's not surprising that many cigarette makers cater to women. They use words like "Slim" and "Light" in brand names. The ads feature very thin women. Sometimes the photos are altered to make the models look even slimmer and taller than they are.

These ads may be having an impact. Recent figures show that, for the first time, *more girls are starting to smoke than boys.* Many girls say that smoking makes them feel less nervous, more sure of themselves, and more easily accepted. Smoking to lose weight is another reason. Smoking may cause some girls to stay thin, but the long-term risks just aren't worth it. There are other ways to stay slim, including the best one of all: eating right.

Boys have their thing, too—so-called "spit" tobacco (snuff and chewing tobacco). The U.S. Surgeon General has reported that these smokeless tobaccos are as "dangerous and deadly as other tobacco products." She noted the "terrific connection between baseball and the use of spit tobacco." (One third of the 700 major league players use it.)

A government report, "Spit Tobacco

and Youth," shows that nearly 20 percent of boys in high school use the stuff in any 30-day period. Boys start chewing as early as age nine. In 1992, 75 percent of the 30,000 new cases of mouth cancer were caused by smoking or chewing. Half of these people will be dead in five years. Spit tobacco delivers nicotine to the body even faster than cigarettes. But because there's no smoke, many people think chewing is safer. *Wrong!*

How Tobacco Causes Problems

The nicotine in cigarette smoke and in chewing tobacco is the one ingredient that creates dependence by forcing your body to demand it. Once the pattern is established, many smokers find they have also developed nicotine *tolerance*. It takes more and more nicotine to satisfy the body's demand.

Nicotine is so poisonous that it is the main ingredient in many insect killers. But it is not the only cause of trouble. Tobacco contains thousands of chemicals. They are released in the smoke and when tobacco is chewed. Carbon monoxide is one poisonous gas. Hydrogen cyanide is another. Snuff contains cadmium (used in car batteries), uranium 235 (found in

Chewing tobacco can cause cancer of the mouth and throat.

nuclear weapons), and polonium 210 (a nuclear waste).

All gases and chemicals condense out of the smoke or the chew to form a brown goo called *tar*. Tar does big-time damage. Tar is a *carcinogen,* a substance that causes cancer.

When you use tobacco, the nicotine and tar reduce the oxygen entering your blood. Oxygen is absorbed into the blood by 300 million air sacs, called *alveoli,* in your lungs. Your heart then pumps oxygen-rich blood to your body through the arteries and blood vessels. The oxygen nourishes your muscles and all the organs in your body, including your brain. When the oxygen in the blood has been used, what is left is carbon dioxide, a waste substance. The veins carry this "used" blood back to the lungs. The alveoli in the lungs exchange the carbon dioxide for fresh oxygen.

The harder you work, play, or exercise, the harder you breathe. This is because your muscles create more carbon dioxide. Your lungs must move faster and your heart must beat faster to give your body more oxygen than usual.

When you breathe in smoke and tar, you irritate your nose, throat, bronchial

46 tubes (windpipe), and lungs. The blood vessels become smaller, so less oxygen is carried into your system, including your brain. Your heart has to pump faster all the time—not just when you exercise—to deliver the oxygen your body needs. And some of the harmful elements in smoke are transferred to your blood along with the oxygen, so they circulate throughout your body.

These things happen right away if you smoke. You can get dizzy or feel sick. If you continue to smoke, the elements in tobacco and tobacco smoke can cause all kinds of health problems. Let's look at them.

How Bad Is Long-Term Smoking?

Very bad. Doctors at a leading medical college recently put together all the facts on the killer effects of smoking. The list may be a bit painful to read, but it shows that smoking can do damage *big time!*

- *Addiction.* We have mentioned this. Breaking off may be hard, but millions of people have done it.
- *Back pain.* Since smoking reduces the oxygen in the blood, the spinal disks do not get enough oxygen.

- *Bladder cancer.* Smoking causes 4,000 cases a year, 40 percent of the total.
- *Breast cancer.* Women smokers are 75 percent more likely to develop this cancer than nonsmokers.
- *Cervical cancer.* Up to 33 percent, or 7,000 cases a year, are caused by smoking. Women who smoke are four times more likely to develop this disease than nonsmokers.
- *Childhood respiratory (breathing) problems.* Kids who live with smokers have six times as many of these infections as other children.
- *Diabetes.* Smoking reduces the level of insulin that is vital to a diabetic. It increases the damage to the small blood vessels in eyes, ears, and feet by depriving them of oxygen.
- *Drug interactions.* Smokers need higher doses for certain medicinal drugs to be effective.
- *Ear infections.* Smokers' children are at greater risk.
- *Emphysema (obstructive lung disease).* Smoking accounts for 85 percent of deaths from this disease. The alveoli in the lungs are permanently damaged by smoke. The more of them

Smoking in a public place affects everyone in the area.

that are damaged, the harder it is to breathe. Emphysema sufferers must breathe oxygen from tanks. Eventually, they die of suffocation.

- *Esophageal (mouth-to-stomach tube) cancer.* Smokers get 80 percent of all cases. The disease kills 15,000 Americans a year.
- *Fires.* Smoking is the number one cause of fires in homes, hotels, and (believe it or not) hospitals.
- *Gastrointestinal (digestive system) cancer.* It is thought that smoking doubles the risk.

- *Heart disease.* Smokers are up to four **49** times as likely to develop heart disease as nonsmokers. Reasons: carbon monoxide and other poisonous gases in tobacco smoke replace oxygen in blood cells. Also, a smoker's heart has to pump harder and faster, placing it under greater strain.
- *Infertility.* Couples at least one of whom smokes are three times more likely to have trouble having babies.
- *Kidney cancer.* Smoking causes 40 percent of all cases.
- *Laryngeal (voice box) cancer.* Smokers of 25 cigarettes or more a day are 25 to 30 times more likely to develop this cancer than nonsmokers.
- *Leukemia (blood cancer).* The tar or condensed smoke in cigarettes contains cancerous chemicals including benzene, a known cause of leukemia.
- *Low birth weight.* Mothers who smoke as few as five cigarettes a day can give birth to underweight babies.
- *Mouth cancer.* Men smokers are 27 times more likely to develop this cancer than nonsmokers. Women smokers (who do not dip snuff or chew tobacco, and who smoke less

50

than men) are six times more likely than nonsmokers.

- *Nutrition.* Smokers tend to eat less of the right foods than nonsmokers.
- *Osteoporosis (breakable bones).* Women smokers reach menopause (the age when they cannot have babies) 5 to 10 years earlier than nonsmokers. For that reason their bones tend to become weaker because of lower estrogen levels. Estrogen is an important hormone that is made in smaller quantities after menopause begins.
- *Pharyngeal (throat) cancer.* The vast majority of the 3,600 who die each year from this disease are smokers.
- *Premature aging.* Smokers develop wrinkles earlier than nonsmokers. Their teeth and fingernails become yellowed from nicotine staining.
- *Recovery.* Smokers take longer to recover from injury or surgery than nonsmokers. They run higher risks of pneumonia. They have longer stays in the hospital.
- *Stroke.* Smoking doubles the risk of this form of paralysis.
- *Tooth loss.* Those who use snuff or

Keeping active helps ensure your body's health.

chew tobacco lose teeth more readily than nonusers.

What About Secondhand Smoke?

The effects of smoking are so bad that they even harm nonsmokers who happen to be around smokers. The Environmental Protection Agency reports that so-called secondhand smoke is very harmful to children who live with smokers. Every year it causes 150,000 to 300,000 cases of breathing infections in babies. It also hurts children with asthma and each year causes up to 26,000 new cases. From

52 | 2,500 to 3,000 nonsmokers die of lung cancer every year. So if you smoke, you don't just hurt yourself.

Think about it!

Why Quit Smoking?

It's very simple. If you quit smoking, your body will start to return to health. If you have smoked for a long time, the healing will take longer. You may not be able to overcome all the effects of smoking, particularly lung problems. But you will reverse a lot of the damage.

How to Quit Smoking

The best way to stop is cold turkey. Just quit! It is also the hardest way, and you may start smoking again. The main problem with going cold turkey is nicotine withdrawal. The symptoms can include craving for a smoke, nervousness, anxiety, restlessness, irritability, mood swings, sleepiness, sleep disturbance, headache, poor concentration, increased appetite, tiredness, and weight gain. The more you have smoked each day, the more likely you are to have one or more of these symptoms. They will pass in time if you can hang in there!

Is there any way besides hanging

tough? Yes, there are several things you can do:

- Find other teens who are trying to stop smoking and form a *support group*, people helping each other to reach a goal. Be sure to invite teens who have quit, so they can be peer counselors and tell you how they did it.
- Consider joining a commercial stop-smoking program. But be careful. They can be very expensive, and success is not guaranteed.

Is hypnotism worth trying? Possibly, if nothing else works. *It must be performed by a professional,* and you must be referred by a doctor or a health clinic.

Hypnotist Sheila Kahn Alper says the method works if you really want to quit. Young people, she says, participate well in hypnosis. Under hypnosis you accept the idea that smoking is a dirty habit. Then your subconscious mind—the part of your brain that directs much of what you do without your being aware of it—helps you stay off tobacco after the hypnotic sessions are over.

54 | *Acupuncture*

Acupuncture has been practiced by the Chinese for 22 centuries. It is now becoming part of our medicine. In acupuncture, superfine needles are painlessly placed under the skin. They stimulate the nerve pathways linking different organs of the body to help relieve pain. If you are willing to be hypnotized, and also believe that acupuncture works, combining them could help you stop smoking. You must be referred by a doctor for acupuncture. Remember, however, there's no guarantee that either hypnotism or acupuncture will work.

You have to *want* to quit. Then your plan stands a better chance of working, whether it's cold turkey or hypnotism.

Two stop-smoking aids are not available to teenagers: nicotine chewing gum and the nicotine patch. They have not been tested on young people.

How about cigarettes with less tar and nicotine? These might help if you didn't smoke more of them to make up for the nicotine you're not getting. *Many smokers who switch to these brands smoke more, not less.*

One technique besides quitting cold turkey works for some: cutting down. The

longer you can put off the first cigarette |
of the day, the better—if you don't make
up for it by smoking more later in the
day. If in time you can put off the "first
puff" from morning to afternoon, then to
evening, and finally to day's end, you can
drop it totally. (Tip: Combine cutting
down with a switch to a low-nicotine
brand of cigarettes.)

Four Things to Remember

1. If you don't start smoking, you won't have to quit.
2. The earlier you start and the more you smoke, the harder it will be to quit.
3. The more you really want to quit, the better your chances.
4. Don't feel ashamed if you start smoking again and have to quit a second, third, or fourth time. *Don't get discouraged!* Eventually, you'll quit for good.

CHAPTER 4

It's Your Choice

Who's in charge here? You are! You and only you can decide how you are going to live your life—and how you are going to treat your body.

You only go around once!

If you drink caffeinated sodas, tea, or coffee, or eat chocolate, be careful to keep the amounts down. If you find you're taking "just one more" cola drink, back off. Set limits for yourself and stay with them. Switch to caffeine-free sodas. Stay away from stay-awake pills!

If you don't smoke, don't start. If you do smoke, stop and ask yourself, "Why am I doing this?"

You are surrounded by images that promote smoking. Luke Perry and Jason

Priestly of "Beverly Hills 90210" were shown with cigarettes in their mouths in a recent national magazine. Old movies make smoking look romantic and macho.

The tobacco companies love this kind of free advertising. They won't admit it, but they want you to smoke. They will not admit the health problems, either, even though the connections between smoking and disease are clear. Instead, they run ads with cool-looking people lighting up—or cartoon characters having fun. They emphasize the macho nature of auto racing and the outdoors and try to relate it to smoking—even though there's no connection. They try to give girls a "slim" message. They spend billions of dollars to promote smoking—far, far more than the antismoking forces are able to spend to stop it.

Remember These Points

1. Caffeine and nicotine are both addictive. Your body will demand them once it tries them.
2. Caffeine and nicotine are both stimulants. They give you a brief "lift," followed by a let-down feeling and a wish for more of the stimulation.

Why quit? Because your body will love you for it.

3. Caffeine and nicotine both cause withdrawal symptoms when you cut down or quit using them.
4. Nicotine is a powerful poison, and the tar that's left over from smoking and chewing tobacco causes many kinds of cancer and other diseases.
5. The later you start and the less you smoke, the easier it will be to quit. (Best idea: Don't start.)
6. Once you quit smoking, you may start up again. *Try to quit again!* Keep hanging in there! Quitting is worth it.

Never forget who's in charge here. You are!

Help List

Associations

American Cancer Society
19 West 56th Street
New York, NY 10019
212-582-2118 ("Stop Smoking" program)

American Heart Association
122 East 42nd Street
New York, NY 10017
212-661-5335

American Lung Association
1740 Broadway
New York, NY 10019
212-315-8700

Nicotine Anonymous
P.O. Box 1468
Baldwin, NY 11510
516-665-0527

Students to Offset Peer Pressure
P.O. Box 103
Hudson, NH 03051-0103

IN CANADA

Canadian Cancer Society
#1702, 77 Bloor Street West
Toronto, ON M5S 3A1
416-961-7223

Canadian Lung Association
#908, 75 Albert Street
Ottawa, ON K1P 5E7
613-237-1208

Glossary
Explaining New Words

acupuncture Ancient Chinese healing art that uses tiny needles to stimulate body nerves and control pain.

addiction When your body demands dangerous things and you are unable to control the demand.

adenosine Body chemical that manages brain activity.

alveoli The air sacs in the lungs; they can be damaged by cigarette smoke.

analeptics Drugs that stimulate the brain.

asthma Illness in which breathing is difficult because the small air passages in the lungs have become narrowed.

bronchial tubes Tubes that carry air from the mouth to the lungs.

bronchitis Irritation of the bronchial tube linings, common among smokers.

cancer One of a group of diseases in which body cells grow out of control.

cold turkey Stopping smoking instantly.

decaffeinated, decaf Coffee, tea, or cola with the caffeine removed.

drug Anything taken into the body that changes the way we feel or behave.

emphysema   Lung disease in which the
air sacs stop working, making breathing difficult.

hypnosis   State of being in a sleeplike
trance, during which you can be told
to do (or not do) certain things, such
as smoke. When you are wakened,
your subconscious mind obeys the
instructions without your being aware
of it.

nicotine   A poison; the addictive substance in tobacco.

peer pressure   Influence exerted on
you by others your age.

receptors   Elements of the brain that,
once influenced by an addictive substance, want more and more of it.

secondhand smoke   Tobacco smoke in
the air.

spit tobacco   Tobacco that is placed in
the mouth instead of being smoked;
also called smokeless tobacco.

tar   The brown substance formed when
tobacco smoke condenses or spit tobacco is moistened.

tolerance   The need for more and more
of a drug to satisfy the craving of addiction.

withdrawal symptoms   The uncom-
fortable effects felt when giving up a
drug (including tobacco).

For Further Reading

Berger, Gilda. *Smoking Not Allowed*. New York: Franklin Watts, 1987.

Gilbert, R. J. *Caffeine: the Most Popular Stimulant*. New York: Chelsea House Publishers, 1986.

Hyde, Margaret. *Addictions: Gambling, Smoking, Cocaine Use and Others*. New York: McGraw-Hill, 1978.

Keyishian, Elizabeth. *Everything You Need to Know about Smoking*, rev. ed. New York: Rosen Publishing Group, 1993.

Perry, Robert. *Focus on Nicotine and Caffeine: the Drug Alert Series*. Chicago: Children's Press, 1990.

Terry, Luther L., M.D., and Horn, Daniel, Ph.D. *To Smoke or Not to Smoke*. New York: Lothrop, Lee and Sheppard Co., 1969.

Ward, Brian R. *Smoking and Health: Life Guides*. New York: Franklin Watts, 1986.

Index

Acknowledgments

We thank Sheila Kahn Alper, hypnotherapist, for information on hypnosis and smoking cessation; Robert Bezilla, Project Director for The George H. Gallup International Institute, Princeton, N.J.; and Lori Morse, Data Base Center, Free Library of Philadelphia, for helping us locate valuable resources.

About the Authors

Richard S. Lee is an advertising writer and free-lance author. Mary Price Lee is a former educator, now a free-lance writer. The Lees also wrote *Drugs and the Media* for the Drug Abuse Prevention Library.

Photo Credits

Cover Photo by Maje Waldo; all other photos by Lauren Piperno.